NIZIĆ / THE EXTREMOLOGY

The EXTREMOLOGY

by Dinko Nizić MD PHD

A Philosophy of
Bounds, Systems,
and Truth

KUNA (Croatian word for *marten*) is a small, inquisitive and clever animal. What is more, the pine marten is a Croatian national symbol. Three semicircles on the tail represent three full decades of Croatian soverenity (1991—2025). Under the *KUNA Label* Medicinarium publishes innovative books from any scientific field, especially biomedicine. The books under this label bring new insights and perspectives into science.

IMPRESSUM

Dinko Nizić
The Extremology

Publisher # Cover & Layout

Medicinarium d.o.o.
Josipa Jurja Strossmayera 28/2
35000 Slavonski Brod, Republika Hrvatska
(Republic of Croatia)

E-mail: medicinarium.publishing@gmail.com

Print

Libri Plureos GmbH
Friedensallee 273
22763 Hamburg, Germany

Proofreading

Dinko Nizić MD PhD

Place of Publication: Slavonski Brod (Republic of Croatia)
Year of Publication: 2025
Sign-off Date: 17-07-2025

CIP data is available in the computer catalogue of the National and University Library in Zagreb under the number 001277800.

ISBN
978-953-46656-0-2

FOREWORD

The Extremology is a poetic philosophy of extreme pragmatic structuralism.

Somewhere on the lines of its ancestral brother, *Tao Te Ching*, and along the principles of French philosopher Gilles Louis René Deleuze on ontological import of difference, this book opens a new chapter on the landscape of thought. It is as poetic as it is pragmatic.

If read carefully,
it is a manifest
of truth.

The Author

CONTENTS

CONTENTS

CONTENTS

The EXTREMOLOGY

BOOK ONE

The Truth

To be is to differ.
Truth is certain.
Truth is in the system.

The Edge

Zero and *one*.
Nothing and *something*.

Where first ends,
the second begins.

Hence,
both are bounded.

Without bounds,
how would nothing end,
and something begin?

Something is direct.
Nothing is indirect.

Hence,
both are present.

Both *are*.

The Nothing

There is nothing.
The presence of absence.

How can absence
be present?

Only if absence
is bounded
as *non-presence*.

If it is bounded,
it is a system,
it is present.

Something is sensed.
Nothing is contrasted.

Hence,
both are present.

Hence,
nihilism is absent.

The Infinity

There is nothing.
There is something.
Both finite.

Infinity
is *infinite*.

No finitude — no bound.
No bound — no system.
No system — no meaning.
No meaning — no truth.

No truth — no presence.

Is infinity
nothing
or
something?

The Relativity

There is nothing.
There is something.
Both finite.

Infinity
is infinite.

The finitude
is relative.

Everything
is relative.

Except — infinity.

And this is so
because
it is infinite.

The Contrast

There is nothing.
There is something.

The beginning of something
is the end of nothing.

The end of something
is the beginning of nothing.

Both bounded,
both systems,
both present.

The beginning of infinity
is the end of finitude.

The end of infinity
is *absent*.

The System

There is nothing.
There is something.

Two halves
of a whole.

A pair
in a system.

Where is
infinity?

The Difference

There is nothing.
There is something.

Two halves
of a whole.

A pair
in a system.

They are *different*,
hence they *are*.

To be
is
to differ.

The Measure

There is nothing.
There is something.

Two halves
of a whole.

A pair
in a system.

They are different,
hence they are.

The *more* different they are,
the *more* they are.

The Magnitude

There is nothing.
There is something.

Two halves
of a whole.

A pair
in a system.

The seed of nothing
is in the seed of something.

The seed of something
is in the seed of nothing.

They are different,
hence they are.

The more different they are,
the more they are.

The more their seeds are *distant*,
the more they are *different*.

The Meaning

There is nothing.
There is something.

Two halves
of a whole.

A pair
in a system.

The more their seeds are distant,
the more they are different.

Yet only
in a whole.

In a *system*.

BOOK TWO

The Lies

To be is to believe.
To be is to repeat.
Truth is uncertain.
Truth is in the infinity.

The Fog

There is nothing.
There is something.

Two halves
of a whole.

A pair
in a system.

In a whole
you *see*.

The unseen whole
becomes probable,
probable becomes uncertain,
uncertain becomes *approximate*.

Approximate
is not truth.

The Belief

There is nothing.
There is something.

Two halves
of a whole.

A pair
in a system.

In a whole
you see.

A seed
they see.

Not a whole
you suggest.

Not a seed
they suppose.

The whole and seed
you *all* see.

The Toss

There is nothing.
There is something.

Two halves
of a whole.

A pair
in a system.

They are different,
hence they are.

How often
should you check?

Ten times?
A hundred times?
A thousand times?

A single check,
a single truth.

The Curve

There is nothing.
There is something.

Two halves
of a whole.

A pair
in a system.

The more their seeds are distant,
the more they are different.

Not their structure.
Not their shape.
Not the symmetry
of their seeds.

A distance — a difference
A difference — a truth.

The Significance

There is nothing.
There is something.

Two halves
of a whole.

A pair
in a system.

The more their seeds are distant,
the more they are different.

Not the smallness of the halves.
Not the greatness of the halves.
Not the benchmark of default.

A distance — a difference
A difference — a truth.

The Prior

There is nothing.
There is something.

Two halves
of a whole.

A pair
in a system.

The more their seeds are distant,
the more they are different.

How much
do you
believe it?

Not much?
Somewhat?
A lot?

A single calculus,
a single truth.

The Uncertainty

There is nothing.
There is something.

Two halves
of a whole.

A pair
in a system.

The more their seeds are distant,
the more they are different.

But —
am I *certain*?

Why are you certain
in your uncertainty?

A single difference,
a single truth.

The Neutrality

There is nothing.
There is something.

Two halves
of a whole.

A pair
in a system.

The more their seeds are distant,
the more they are different.

But —
why watching?
Why touching?

By watching,
the truth appears.
By touching,
the truth reaffirms.

You are *not* the truth.

A single difference,
a single truth.

The Model

There is nothing.
There is something.

Two halves
of a whole.

A pair
in a system.

The more their seeds are distant,
the more they are different.

But —
why model?

Do you not see it?
Do you see it?

The system *is* the model.

A single difference,
a single truth.

The Complexity

There is nothing.
There is something.

Two halves
of a whole.

A pair
in a system.

The more their seeds are distant,
the more they are different.

But —
why so simple?

Truth is simple,
but the road to the truth
is complex.

The road *is not* the truth.
The *truth* is the truth.

And truth is
the difference.

BOOK THREE

The Way

Truth is a difference.
Truth is a calculus.
Truth is a formula.

The Trimm

There is nothing.
There is something.

Two halves
of a whole.

A pair
in a system.

A pair
deformed and
distorted.

Cut off the ends!
Liberate the middle!

New ends,
old percentiles.

The Seed

There is nothing.
There is something.

Two halves
of a whole.

A pair
in a system.

Each
with a middle.

The seed is
in the middle's middle.

The seed is
the edges
by half.

The Summ

There is nothing.
There is something.

Two halves
of a whole.

A pair
in a system.

And what is
a system
but the edges
summed?

The Formula

There is nothing.
There is something.

Two halves
of a whole.

A pair
in a system.

They are different,
hence they are.

Different as much
as the seed's distance
by the edges' summ.

This is the formula
of *meaning*.

The Mirror

There is nothing.
There is something.

Two halves
of a whole.

A pair
in a system.

Nothing on the left.
Something on the right.

Something on the right.
Nothing on the left.

And the system
around.

The Quiet

There is nothing.
There is something.

Two halves
of a whole.

A pair
in a system.

Their seeds
too close.

No difference — no pair.
No pair — no truth.

The Doom

There is nothing.
There is something.

Two halves
of a whole.

A pair
in a system.

Their seeds
too apart.

They left the system.

And the system
left presence.

The Untruthness

There is nothing.
There is something.

Two halves
of a whole.

A pair
in a system.

Their seeds too close —
the quiet untruthness.

Their seeds too apart —
the loud untruthness.

Their seeds at a distance —
the loud truth.

The Scale

There is nothing.
There is something.

Two halves
of a whole.

A pair
in a system.

Large or small
the edges change,
yet the meaning
remains.

The truth is not the size.
The truth is the seed's distance
by the edges' summ.

Truth is
the *ratio*.

The Everywhere

There is nothing.
There is something.

Two halves
of a whole.

A pair
in a system.

A sound.
An image.

The system — the meaning.
The meaning — the difference.
The difference — the truth.

Truth is
in the system.

And system is
everywhere.

BOOK FOUR

The Psaltyr

Truth is in the system.

The Cure

There is nothing.
There is something.

Two halves
of a whole.

A pair
in a system.

The more different they are,
the more they are.

Small group.
Perhaps cure.
How much it helps?

Truth is
in the system.

The Word

There is nothing.
There is something.

Two halves
of a whole.

A pair
in a system.

The more different they are,
the more they are.

A fragmented scroll.
An ancient language.
What does it say?

Truth is
in the system.

The Image

There is nothing.
There is something.

Two halves
of a whole.

A pair
in a system.

The more different they are,
the more they are.

A sacred image.
A blurred reflection.
What does it show?

Truth is
in the system.

The Sound

There is nothing.
There is something.

Two halves
of a whole.

A pair
in a system.

The more different they are,
the more they are.

The signal.
The noise.
What is the sound?

Truth is
in the system.

The Ethics

There is nothing.
There is something.

Two halves
of a whole.

A pair
in a system.

The more different they are,
the more they are.

The moral.
The immoral.
Where is the righteous truth?

Truth is
in the system.

The Road

There is nothing.
There is something.

Two halves
of a whole.

A pair
in a system.

The more different they are,
the more they are.

The road to the left.
The road to the right.
Which road to take?

Truth is
in the system.

The Time

There is nothing.
There is something.

Two halves
of a whole.

A pair
in a system.

The more different they are,
the more they are.

The before.
The after.
What has changed?

Truth is
in the system.

The Reality

There is nothing.
There is something.

Two halves
of a whole.

A pair
in a system.

The more different they are,
the more they are.

The real.
The unreal.
Are they real?

Truth is
in the system.

The Feel

There is nothing.
There is something.

Two halves
of a whole.

A pair
in a system.

The more different they are,
the more they are.

The love.
The hate.
Where is the truth?

Truth is
in the system.

The Life

There is nothing.
There is something.

Two halves
of a whole.

A pair
in a system.

The more different they are,
the more they are.

To be or
not to be?
That is the question.

The answer is
in the system.

And the answer
is
truth.